NATURE'S RULE BREAKERS

CREATURES THAT DON'T FIT IN

JESSICA FRIES-GAITHER

Millbrook Press / Minneapolis

People *try* to organize nature into categories.

But nature doesn't always cooperate.

Sometimes it breaks the rules.

PACIFIC SWALLOW

MEET THE IN-BETWEENERS!

LITTLE BROWN BAT

They're bold animals that are *BOTH*, notable creatures that are *NEITHER*, and fantastic living things that don't quite *FIT*. They were born to STAND OUT!

EARLY BIRD OR NIGHT OWL?

Diurnal animals are active during the day, and nocturnal creatures are awake at night. But what if neither feels right?

Meet the **EURASIAN EAGLE-OWL.**
Unlike some other owls, it isn't nocturnal. This feathered flier is crepuscular (kri-PUH-skyuh-luhr), swooping silently through the sky at dawn or dusk. That's when its prey is most active too.

And what if an animal's sleeping habits change?

The BOBCAT, the RED-FRONTED LEMUR, and the RIVER OTTER are unpredictable. Scientists call them cathemeral (ca-THEM-er-uhl) animals. Active at any time, they don't follow a set sleeping pattern.

WARM-BLOODED OR COLD-BLOODED?

Many people think animals are *either* warm-blooded or cold-blooded. Not so fast . . .

The **GREAT WHITE SHARK** doesn't make enough heat to keep a constant body temperature as warm-blooded animals do. And it doesn't completely depend on its environment to stay warm as cold-blooded animals do. This ferocious fish makes *some* heat for itself, but its temperature can still change. One benefit of staying warmer than the ocean water is that the shark can swim faster than its cold-blooded prey.

The ECHIDNA (eh-KID-nuh), TUNA, and LEATHERBACK TURTLE aren't warm-blooded or cold-blooded either. Whether on land or in the water, their body temperatures fluctuate but still stay warmer than their surroundings. These in-betweeners don't get sluggish in low temperatures as cold-blooded creatures do. And they don't need to eat as much as warm-blooded creatures do to maintain their body heat.

ARE PLANTS OR ANIMALS ON THE MENU?

People may believe that herbivores *only* graze on plants and that carnivores *only* eat other animals. But it's not so simple.

Meet the **WHITE-TAILED DEER**.
Deer mostly eat plants such as twigs, leaves, and grass, but sometimes only meat will satisfy these hungry herbivores. They have been known to gobble hatchlings straight from a nest!

Other animals mix up their diets too.

RETICULATED GIRAFFES chew on bones for minerals they can't get from plants. HIPPOS indulge in the occasional meaty snack. And ALLIGATORS and CROCODILES sometimes eat fruit!

FRESHWATER OR SALTWATER HABITATS?

YELLOWSTONE CUTTHROAT TROUT

EMPEROR ANGELFISH

Most types of fish live in either fresh water or salt water. Their bodies are designed to maintain the right amount of salt. Fish that live in salt water must get rid of extra salt. Freshwater fish need to hold on to as much salt as they can. But what if both places feel like home?

Meet the **SOCKEYE SALMON**.
Its life cycle takes it from fresh water to salt water and back again. A baby salmon hatches from an egg nestled in the gravelly bottom of a freshwater stream or lake. It spends one to three years in its freshwater home before swimming out to the salty sea. The ocean then becomes home for several years as it grows. Finally, when it's ready to spawn, the adult salmon journeys back to the same place where it hatched.

The **AMERICAN EEL** also moves between salt and fresh water. A long, leaf-shaped, transparent eel larva hatches from an egg floating in the Sargasso Sea of the Atlantic Ocean. Ocean currents carry the developing eel toward the eastern coast of the United States where rivers meet the sea. Then the eel swims up the river to a freshwater home. Once fully grown, it migrates back to the ocean to spawn.

MALE OR FEMALE?

People often think living things are *either* male or female. Do they have to be just one?

Meet the GARDEN SNAIL.
It has both male and female parts. It's known as a hermaphrodite (huhr-MAF-ruh-dite). It can mate with any other snail of its kind. When snails mate, both of them can lay eggs. One snail can also reproduce without a mate.

Some animals even change from one sex to the other. While growing inside their eggs, male BEARDED DRAGONS will become female if the temperature is too hot.

And all CLOWN FISH hatch as males. When they grow up, the biggest fish in the school becomes female and is the leader of the group. When she dies, the next largest male becomes female and takes her place.

LUNGS OR GILLS?

Many animals get the oxygen they need with either lungs *or* gills. Most creatures that live in water have gills. Animals that live on land breathe through their lungs. But why not use both?

Meet the AUSTRALIAN LUNGFISH.
It has gills *and* a lung. The lungfish lives in pools of water near rivers. Most of the time, it uses its gills to take in oxygen from the water. Sometimes the pools dry up, but the lungfish can breathe in air with its single lung when the water levels are too low. Having both sure comes in handy!

The **RED SALAMANDER** lacks lungs *and* goes without gills. This amphibian doesn't need them. It breathes through its skin!

PRODUCE FOOD OR JUST EAT IT?

Plants make their own food using sunlight, water, and air. Animals have to eat other living things to get their energy. That's almost always true, but . . .

Meet the **SACOGLOSSAN SEA SLUG**.
It can make its own food—with a little help. It normally feeds on algae but doesn't fully digest them. Instead, it stores the algae's chloroplasts in its own body. It uses them to make food when algae are scarce. Just call it a solar-powered sea slug!

DEAD OR ALIVE?

OPPOSSUM

Living things usually die when their habitats become too hot, too cold, too wet, or too dry. But what if they just *look* like they're dead?

Meet the **WOOD FROG**.
It survives the winter by freezing completely solid. Its heart stops beating, and it stops breathing. Still, it's not dead. Incredibly, the frog will thaw from the inside out when the weather warms up. First, the heart starts beating. Next, the brain begins to work again. Finally, its legs move. It's as good as new!

TARDIGRADES can appear to be dead for a LONG time. These tough, teeny-tiny creatures live all over the world: in pools on the Himalayan mountains, on damp moss growing in North American forests, and even in the Arctic Ocean. When conditions aren't right, they curl up and go into *tun*. A tardigrade tun is a dried-out ball that *seems* dead. But years—or even decades—later, they can become active again.

Do these animals mean to break the rules?
Not at all.
It's simply who they are.

DUCK-BILLED PLATYPUS

Our categories don't fit *them*.
Standing OUT is IN their nature.

AFRICAN GROUND PANGOLIN

WHY DO WE HAVE IN-BETWEENERS, ANYWAY?

It's all because people, especially scientists, like to categorize things, or put them into groups based on characteristics they have in common. Grouping animals and plants can help us understand how different living things are related to one another, and it helps us talk about them with others. Categories make sense to our brains and make understanding the world easier. Some of the categories we use to group animals include warm- and cold-blooded, nocturnal and diurnal, and groups such as mammals, birds, reptiles, amphibians, insects, and fish.

The problem is that nature is too wild and wonderful to ever perfectly fit into categories. For every group or category we create, there are always exceptions that don't fit. Any time we try to divide living things into neat groups, we end up with in-betweeners. But isn't the world so much more interesting that way?

There are many, many more examples of in-betweeners than could fit in the pages of this book! And there have been throughout Earth's history.

EXTINCT IN-BETWEENERS

Artist rendering of a Tiktaalik

TIKTAALIK (tick-TALL-ik) was an ancient fish that lived about 375 million years ago. Even though it lived in the water, it had strong back limbs that may have helped it crawl as well as swim. Because of these back legs and other features such as a movable neck and a crocodile-like head, scientists believe that Tiktaalik was an ancestor of land animals.

Ambulocetus natans

AMBULOCETUS NATANS, or "walking whale," lived about forty-nine million years ago in shallow, brackish water in what is now northern Pakistan. *Ambulocetus* had short, strong legs and flipper-like hind feet, which helped scientists understand that whales are closely related to hippos.

IN-BETWEENERS OF THE PLANT KINGDOM

Animals aren't the only living things that can be in-betweeners. Some plants break the rules as well.

Pitcher plant

The **PITCHER PLANT** and the **VENUS FLYTRAP** are two examples of carnivorous plants, or plants that eat insects. Most plants make their own food using sunlight, water, and air. Carnivorous plants do this as well, but they need insects to give them extra nutrients they can't get from the soil in their habitats.

Venus flytrap

Larches

LARCHES are a group of conifer trees with needles and seeds that grow in cones. But they are unlike pine trees for one big reason: their needles turn yellow and fall off in the fall, while pine trees are evergreen and keep their needles year-round.

INDIVIDUAL IN-BETWEENERS

Postman butterfly

All the in-betweeners described in this book are species, or a specific kind of animal or plant. But sometimes, individual animals have traits that make them in-betweeners too! Birds, butterflies, lobsters, crabs, and insects can be born half male and half female. These animals often have an unusual combination of colors and markings, like this **POSTMAN BUTTERFLY** with one set of female wings and one set of male wings.

GLOSSARY

AMPHIBIAN: a member of a group of cold-blooded animals with moist skin that lives part of its life in water and part on land

CARNIVORE: an animal that eats other animals

CATEGORY: a group of objects or items that have traits in common

CATHEMERAL: awake and active at any time of day or night

CHLOROPLAST: a part of a plant that makes food using sunlight, water, and air

COLD-BLOODED: an animal that doesn't produce heat and relies on its surroundings to keep itself warm

CREPUSCULAR: awake and active at dawn or dusk

DIURNAL: awake and active during the daytime

HERBIVORE: an animal that eats plants

HERMAPHRODITE: an animal with both male and female body parts

LARVA: the young stage of an insect, amphibian, or fish

MIGRATE: to move from one place to another

MINERAL: pure nonliving solid materials formed by the earth

NOCTURNAL: awake and active at night

SPAWN: to mate

TARDIGRADE: a microscopic, eight-legged, segmented animal related to arthropods such as insects and spiders

WARM-BLOODED: an animal that controls its body temperature by producing heat

FURTHER READING

You can find the complete bibliography of sources consulted at https://www.jessicafriesgaither.com/nature-s-rule-breakers

BOOKS

Broom, Jenny. *Animalium: Welcome to the Museum*. London: Big Picture, 2014.

Levine, Sara. *The Animals Would Not Sleep!* Watertown, MA: Charlesbridge, 2020.

Levine, Sara. *Tooth by Tooth: Comparing Fangs, Tusks, and Chompers.* Minneapolis: Millbrook Press, 2016.

Stewart, Melissa. *Summertime Sleepers: Animals That Estivate*. Watertown, MA: Charlesbridge, 2021.

WEBSITES

National Geographic Kids: Animals
https://kids.nationalgeographic.com/animals

PBS Kids Creaturepedia
https://pbskids.org/wildkratts/creaturepedia/

Ranger Rick: Animals
https://rangerrick.org/animals/

For Marshall and Claire. May you always delight in wild and wonderful nature. —J.F.-G.

Millbrook Press™
An imprint of Lerner Publishing Group, Inc.
241 First Avenue North
Minneapolis, MN 55401 USA

For reading levels and more information, look up this title at www.lernerbooks.com.

Image credits: © copyright kengoh8888/Getty Images, p. 2; Mary Ann McDonald/Getty Images, p. 3; Gary Carter/Getty Images, p. 4 (left); ca2hill/Getty Images, p. 4 (right); Izumi T/Getty Images, p. 5; John Conrad/Getty Images, p. 6 (left); mirecca/Getty Images, p. 6 (upper right); opulent-images/Getty Images, p. 6 (lower right); Layne Kennedy/Getty Images, p. 7 (left); Alan Tunnicliffe Photography/Getty Images, p. 7 (right); Martin Heyn/Getty Images, p. 8; Michael Barrett/EyeEm/Getty Images, p. 9 (top left); Rodrigo Friscione/Getty Images, p. 9 (bottom left); Mark Meredith/Getty Images, p. 9 (right); annick vanderschelden photography/Getty Images, p. 10 (left); Westend61/Getty Images, p. 10 (right); Cameron Raulston/500px/Getty Images, p. 11; Gerald Corsi/Getty Images, p. 12 (left); AlexDibnah/Shutterstock, p. 12 (top right); Moelyn Photos/Getty Images, p. 12 (bottom right); mlharing/Getty Images, p. 13 (left); Georgette Douwma/Getty Images, p. 13 (right); mlharing/Getty Images, p. 14; Michael Wood/Stocktrek Images/Getty Images, p. 15; Papillo/Alamy Stock Photo, p. 16; Dudley Simpson/Shutterstock, p. 17; tiaramaio/Getty Images, p. 18; by Roberto Peradotto/Getty Images, p. 19 (left); bthompso2001/Getty Images, p. 19 (right); Tom McHugh/Science Source, p. 20; JasonOndreicka/Getty Images, pp. 21, 29 (background); cbimages/Alamy Stock Photo, p. 22; Christina Shaskus/EyeEm/Getty Images, p. 23; Ted Kinsman/Science Source, p. 24; STEVE GSCHMEISSNER/SCIENCE PHOTO LIBRARY/Getty Images, p. 25; Robin Smith/Getty Images, p. 26; CarlFourie/Getty Images, p. 27; Danny Ye/Shutterstock, p. 28 (bottom); Dotted Yeti/Shutterstock, p. 28 (top); Kelly Marken/Shutterstock, p. 29 (top left); Vasilyev Alexandr/Shutterstock, p. 29 (middle left); Andreas Häuslbetz/Getty Images, p. 29 (right); James H. Robinson/Science Source, p. 29 (bottom left); Kevin Schafer/Getty Images, p. 31. Design element: Marcia Straub/Getty Images. Cover: JasonOndreicka/Getty Images; Lillian King/Getty Images; Joe McDonald/Getty Images; mirecca/Getty Images.

Designed by Viet Chu.

Main body text set in CCDaveGibbonsLower.
Typeface provided by Comicraft.

Library of Congress Cataloging-in-Publication Data

Names: Fries-Gaither, Jessica, 1977- author.
Title: Nature's rule breakers: creatures that don't fit in / Jessica Fries-Gaither.
Description: Minneapolis, MN, USA: Millbrook Press, an imprint of Lerner Publishing Group, Inc., [2024] I Includes bibliographical references. I Audience: Ages 5–9 I Audience: Grades K–1 I Summary: "Scientists often organize animals into categories based on similar traits and behaviors. But what if an animal doesn't fit in? Author Jessica Fries-Gaither introduces eight common categories and the animals that break those rules"– Provided by publisher.
Identifiers: LCCN 2022049801 (print) I LCCN 2022049802 (ebook) I ISBN 9781728477206 (lib. bdg.) I ISBN 9798765602195 (eb pdf)
Subjects: LCSH: Animal behavior–Juvenile literature. I Animals–Juvenile literature.
Classification: LCC QL751.5 .F74 2024 (print) I LCC QL751.5 (ebook) I DDC 591.5–dc23/eng/20221116

LC record available at https://lccn.loc.gov/2022049801
LC ebook record available at https://lccn.loc.gov/2022049802

Manufactured in the United States of America
1-52356-50713-3/13/2023